CLIMATE CHANGE

STOPPING CLIMATE CHANGE

BY MARTHA LONDON

CONTENT CONSULTANT
Barry Rabe, PhD
Professor of Public Policy and Environmental Policy
Gerald R. Ford School of Public Policy
University of Michigan

Cover image: Using renewable energy, such as solar energy and wind energy, helps slow climate change.

Core Library

An Imprint of Abdo Publishing
abdobooks.com

abdobooks.com

Published by Abdo Publishing, a division of ABDO, PO Box 398166, Minneapolis, Minnesota 55439. Copyright © 2021 by Abdo Consulting Group, Inc. International copyrights reserved in all countries. No part of this book may be reproduced in any form without written permission from the publisher. Core Library™ is a trademark and logo of Abdo Publishing.

Printed in the United States of America, North Mankato, Minnesota
082020
012021

THIS BOOK CONTAINS
RECYCLED MATERIALS

Cover Photo: Shutterstock Images
Interior Photos: iStockphoto, 4–5, 24–25, 28, 37, 43; Red Line Editorial, 8, 39; Chris Boswell/iStockphoto, 9; SS Studio Photography/Shutterstock Images, 11; Mark Elias/Bloomberg/Getty Images, 14–15; Shutterstock Images, 18, 22, 45; Vladyslav Starozhylov/Shutterstock Images, 21; Andrew Bertuleit/iStockphoto, 26; Sean Pavone/iStockphoto, 32–33

Editor: Marie Pearson
Series Designer: Katharine Hale

Library of Congress Control Number: 2019954179

Publisher's Cataloging-in-Publication Data

Names: London, Martha, author
Title: Stopping climate change / by Martha London
Description: Minneapolis, Minnesota : Abdo Publishing, 2021 | Series: Climate change | Includes online resources and index
Identifiers: ISBN 9781532192753 (lib. bdg.) | ISBN 9781644944288 (pbk.) | ISBN 9781098210656 (ebook)
Subjects: LCSH: Environmental protection--Citizen participation--Juvenile literature. | Clean energy industries--Juvenile literature. | Green movement--Juvenile literature. | Climatology--Research --Juvenile literature. | Pollution prevention--Citizen participation--Juvenile literature.
Classification: DDC 363.738--dc23

CONTENTS

KEEPING EARTH HEALTHY

A woman turns on her car radio. The show gives her a traffic update. She is stuck in a backup again. It is going to take her an extra 20 minutes to get to work. Her car idles in a line of other vehicles. She watches puffs of exhaust come out of the tail pipes. The woman knows that her car adds carbon dioxide to the atmosphere. It adds to the greenhouse gases that make Earth's climate warmer. The woman decides to do something to help the planet.

Reducing the amount of time spent driving helps lower carbon emissions that are harmful to the climate.

The next week, she takes a train to work. It is part of a big public transportation system in her city. The woman spends more time getting to work. However, she does not have to worry about traffic. She spends the train ride reading a book. It is a short walk from the train station to her office.

On the weekend, she rides her bike to get groceries. The woman feels good. She is cutting out some of the carbon emissions that contribute to climate change. She begins to think of other ways to live greener and help other people cut back on their emissions too. To live greener means to reduce how much carbon dioxide and other greenhouse gases one releases into the atmosphere.

WHAT IS CLIMATE CHANGE?

Climate change is the effects caused by rising levels of greenhouse gases in the atmosphere. Earth's climate has natural changes. However, today's climate change is caused mostly by human actions. Humans burn fossil fuels such as oil and coal. Burning these fuels releases gases into the air. Some gases, such as carbon dioxide, stay in the air for a long time.

The gases in the air trap the sun's heat and reflect it back to Earth's surface. This is called the greenhouse effect. The greenhouse effect helps keep Earth warm. However, when too many gases stay in the air, they trap more heat. Earth gets warmer.

A warming Earth can have some short-term benefits. For example, growing seasons for crops may be longer. However, climate change also

NOT JUST CARBON DIOXIDE

Carbon dioxide is not the only greenhouse gas. Greenhouse gases also include nitrous oxide, fluorinated gases, methane, and water vapor. Nitrous oxide comes from farming. Fluorinated gases are produced by refrigerators. Methane warms Earth more than carbon dioxide. But methane does not stay in the air as long. Sources of methane include livestock, crop farming, fossil fuels, and landfills.

CARBON DIOXIDE EMISSIONS BY SOURCE, AUGUST 2019

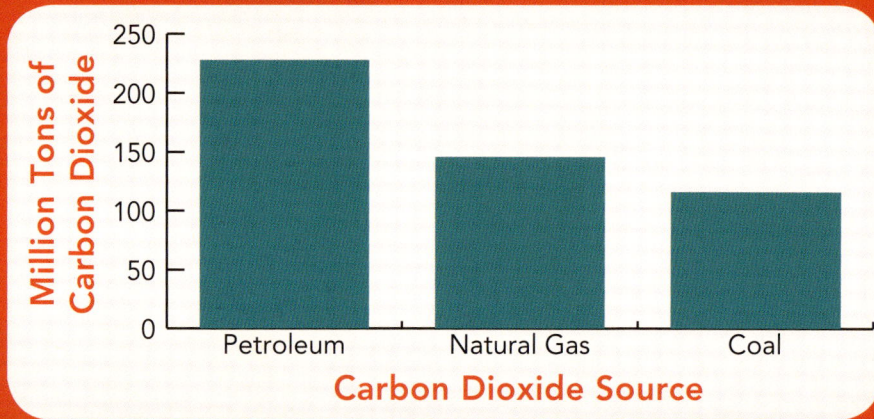

This graph shows the three main sources of carbon dioxide emissions created from generating energy in the United States in August 2019. The measure of 1 ton is equal to 0.9 metric tons. What do you notice about the graph? How does this graph help you better understand the text?

makes life on Earth much harder. Some people and animals have already begun to experience these effects.

WIDESPREAD EFFECTS

All regions of the world are affected by climate change. Ice in the Arctic and Antarctic is melting. Sea levels are rising. Additionally, the ocean is getting warmer. As the ice melts and the ocean warms, ecosystems

Climate change can lead to more rain and snow in some areas, causing flooding, as it did in the Midwest in 2019.

are changing. Plants and animals are not able to adapt quickly enough.

As climate change worsens, so will storms. Scientists believe storms will become more severe in the future. Severe weather can flood cities and flatten homes.

PERSPECTIVES

THE PARIS CLIMATE AGREEMENT

The Paris Climate Agreement is a global plan to stop climate change. Countries agreed to reduce emissions. António Guterres, the United Nations Secretary General, gave a speech in 2017. He said it was important for nations to work together. It is not possible to stop climate change alone. He said, "If any government doubts the global will and need for this accord, that is reason for all others to unite even stronger and stay the course."

While some areas will receive more rain, others will get less. Droughts will become more common in some areas. As regions dry out, wildfires have a higher chance of starting. Today, fires in some regions last longer and are hotter than in previous years. There is little moisture to stop the fires from spreading.

A TEAM EFFORT

The dangers of climate change are real. But scientists believe it is possible to slow and eventually stop

Drought made Australian wildfires from 2019 to 2020 hard to put out. The fires made the air dangerous to breathe.

climate change. People and countries need to work together. It will not be easy.

Slowing climate change will take more than reducing greenhouse gas emissions. Scientists also need to find a way to remove extra carbon from the air. If people do not remove the extra carbon, the harmful effects of climate change will continue.

There are many ways to remove carbon dioxide. But scientists are working to find a way to remove large amounts quickly. Carbon-removal systems can pull carbon from the air. Companies can use the systems in their factories. They can remove carbon before it reaches the atmosphere. But more research needs to be done on these technologies to know how effective they will be. There are other ways that scientists know will help slow climate change. One important way is to reduce people's carbon emissions.

STRAIGHT TO THE
SOURCE

Katharine Hayhoe is a scientist. She studies climate change. In a 2019 interview, she explained how public opinion about the climate is changing.

We haven't yet reached the tipping point to motivate sufficient action. But there has been a change. Ten years ago, few people felt personally affected by climate change. It seemed very distant. [In 2019], most people can point to a specific way climate affects their daily lives. This is important because the three key steps to action are accepting that climate change is real, [recognizing] it affects us, and being motivated to do something to fix it. Opinion polls in the US show 70% of people agree the climate is changing, but a majority still say it won't affect them.

Source: Jonathan Watts. "Katharine Hayhoe: 'A Thermometer Is Not Liberal or Conservative.'" *Guardian*, 6 Jan. 2019, theguardian.com. Accessed 16 Dec. 2019.

BACK IT UP

The author of this passage is using evidence to support a point. Write a paragraph describing the point the author is making. Then write down two or three pieces of evidence the author uses to make the point.

CARBON FOOTPRINTS

One way to help stop climate change is for people to reduce their carbon footprint. A carbon footprint is the amount of carbon produced while making a product or doing an activity. Every person, country, and company has a carbon footprint. Many people use fossil fuels to power vehicles. Energy companies burn fossil fuels to create electricity. The amount of energy a person uses adds to that person's footprint.

Online purchases impact a person's carbon footprint. Packages may have to travel thousands of miles to be delivered.

The smaller the footprint, the better it is for Earth's climate. Eventually, scientists want all carbon emissions to be cancelled out by other carbon-clearing actions. Then the amount of carbon in the air will no longer rise.

CARBON BY COUNTRY

Some people have larger footprints than others. In 2016 the average person in the world produced 4.8 tons (4.35 metric tons) of carbon dioxide each year. Saudi Arabia and Australia had the highest carbon footprints per person of any nation that year. The average person in those countries created more than 17 tons (16 metric tons) of carbon dioxide each year. The United States followed with 16.5 tons (15 metric tons) of carbon dioxide per person. Canada was just below the United States.

People in these countries produce more emissions for a few reasons. One is that the United States, Canada, and Australia have easy access to fossil fuels. Mining coal and drilling for oil and gas produce a lot of carbon dioxide. These nations are also large. People often need to travel some distances to get to work or visit friends. Many regions are rural. Such places have no reliable sources of public transportation. Most people rely on single-person vehicles.

THE POWER OF A LIGHT BULB

There are many types of light bulbs. Some use less energy than others. The most efficient are light-emitting diode (LED) bulbs. LED bulbs last longer than traditional bulbs. They also use 75 percent less energy. If most Americans started using LEDs, it could save the same amount of energy that 44 power plants produce in a year. This would also save $30 billion total in energy costs.

Additionally, carbon is not the only issue. Many people eat meat. Livestock produce the greenhouse gas methane. Methane comes from livestock's waste. Approximately 16 percent of global emissions come from methane. Some people choose not to eat any meat or dairy in order to reduce these emissions. But cutting meat and dairy completely is not

Cows produce significant amounts of methane. Eating less meat will lower the need for livestock and reduce methane emissions.

necessarily healthier for the environment. Some milk alternatives such as almond milk require large amounts of water to produce. It is important that people not waste food or eat too much of one thing.

It is important for people to know their carbon footprint. However, companies produce much more carbon dioxide than any single person. In 2015, 25 companies created 40 billion tons of carbon dioxide. They also produced other greenhouse gases. From 1854 to 2015, just 100 companies produced 71 percent of all greenhouse gas emissions.

THE CARBON TAX

In order to stop climate change, companies need to change their practices. Some countries charge companies taxes for the carbon they create. Carbon taxes encourage companies to choose greener energy sources and use less fossil fuels. Then they can avoid higher taxes. However, being taxed or using alternative energy sources can cost more money. As a result, the company may raise the prices of goods. Consumers have to pay more for those goods. The use of taxes to change consumer habits is not new. For example, the US government taxes tobacco. States also have cigarette taxes. Due to tax increases, tobacco use dramatically decreased.

PERSPECTIVES

MEATLESS MONDAY

Scientists say that cutting out meat one day per week can have a big effect in reducing greenhouse gas emissions. Students in New York City public schools started Meatless Mondays in 2019. This change resulted in 79 million meatless meals each year. New York City's director of the Mayor's Office of Sustainability, Mark Chambers, said in a press release, "Meatless Mondays will introduce hundreds of thousands of young New Yorkers to the idea that small changes in their diet can create larger changes for their health and the health of our planet."

Some countries tax people directly for carbon. For example, Canada began a carbon tax in some provinces and territories in 2019. The taxes started at $10 per ton of carbon dioxide. Each year, the rate rose a little. The tax was scheduled to stop rising in 2022 at $50 per ton. Products such as gasoline are taxed. Carbon taxes cause people to avoid actions that are taxed. British Columbia had a carbon tax before

Carbon taxes can affect the prices of products such as gasoline.

Canada's federal tax. The province's emissions dropped 3.7 percent in ten years, even while its population grew. Due to Canada's tax, carbon taxes are gaining interest in the United States.

Making changes to homes can reduce carbon emissions. But not everyone has the money for these changes. The government helps people. It gives people incentives for producing less carbon. The incentives include giving money back to people for making green choices. For example, people may buy an electric vehicle. However, these incentives do not cover the

Solar panels have been installed in Nevada's desert as a source of renewable energy.

whole cost. People still have to pay thousands of dollars to make these changes.

Carbon taxes can raise a lot of money. Governments have many options for spending that money. It can be used to plant trees. Some countries choose to cut other taxes. In some US states, money raised goes back to installing more renewable energy sources. Carbon taxes are just one way governments persuade people to lower emissions. But stopping climate change will take more than a tax. It requires people to change how cities are designed.

EXPLORE ONLINE

Chapter Two discusses carbon footprints. The website below explores ways people can reduce their carbon footprints. What information does the website give about carbon and emissions? How is the information from the website the same as the information in Chapter Two? What new information did you learn from the website?

NASA CLIMATE KIDS: HOW TO HELP

abdocorelibrary.com/stopping-climate-change

CITY REDESIGN

Cities are designed carefully. They must fit many people and cars. Cities are often near major highways. They are centers of activity containing roads, vehicles, parks, businesses, and homes.

Cities have been designed similarly for many years. Designers often structure them around cars instead of pedestrians. People in cities don't have to travel as far for work and shopping, so they produce less carbon dioxide than those in suburban and rural areas.

Skyscrapers allow people and businesses to be located close together, so people in cities spend less time driving.

Central Park in New York City is an iconic green space.

Still, many people live in cities. Combined, they produce a lot of carbon dioxide. Climate scientists say cities need to be designed differently. Different design choices could reduce carbon emissions.

GREEN SPACES

Green spaces are areas with plants. They are important in cities. Plants take in carbon dioxide. They give off oxygen. They improve air quality. Green spaces reduce the carbon footprint of a city.

Rooftop gardens are one location for green spaces. There are many benefits to rooftop gardens. They absorb carbon dioxide. Rooftop gardens also lower a building's energy needs. During hot summers, the gardens keep buildings cooler.

Additionally, rooftop gardens hold water. Cities have a high risk of flooding. So much of the ground is paved that rain cannot soak into the soil. During the summer, rooftop gardens hold up to 90 percent of the rain that falls. This lessens the likelihood of flooding.

Vertical forests are a way to bring more plant life into a city. Vertical forests are buildings with many plants growing from their walls. Building a city requires clearing thousands of trees from an area. Vertical forests can help reforest a city.

There are two experimental vertical forests in Milan, Italy. The tall buildings hold 900 trees and more than 20,000 other plants. Designers used many types of plants. They chose a variety of shrubs, flowers, and trees. Birds and insects make their homes in the vertical forests. The plants absorb dust and carbon dioxide. They lower the energy costs for the buildings.

The vertical forests in Milan help make the air in the city cleaner.

PUBLIC TRANSIT

Cars are the least efficient form of transportation in a city. They are the main source of greenhouse gas emissions from transportation. Also, cars may hit bikers or pedestrians. Public transportation can help all of these issues in a city. When many people use public transportation, they don't create as much carbon.

There are even more ways to reduce emissions. One option is using hybrid buses. Some of these buses use a combination of fossil fuels and rechargeable batteries. Regular buses use only fossil fuels. In 2019 the city of Nashua, New Hampshire, replaced two of its regular buses with hybrid buses. The city expected to save 4,000 gallons (15,100 L) of fuel each year. As a result, the buses will save 50 tons (45 metric tons) of carbon. But new buses

COMMUNITY GARDENS

Community gardens are vegetable and flower gardens that local residents use. Community gardens fight climate change in several ways. Gardeners compost kitchen scraps from home to make new soil. The scraps do not end up in landfills. Landfills produce methane, a greenhouse gas. Community gardens absorb carbon dioxide. Additionally, the gardens give people access to fresh produce. People do not have to rely on processed foods. Processed foods have a larger carbon footprint. They have to go through more steps before reaching the consumer. Each step a food goes through produces carbon.

are expensive. A new hybrid bus costs approximately $680,000.

Electric streetcars, trains, and subway systems are other options. However, depending on where the electricity comes from, these options could still cause emissions. Streetcars make frequent stops. They don't travel very quickly. Subways travel faster. The New York City subway system has 659 miles (1,061 km) of track. The subway runs through the city.

Trains and streetcars are expensive to install.

PERSPECTIVES

CHANGING THE WAY WE DESIGN

Helen Clark is a former New Zealand prime minister. In 2018 she gave a speech at a conference for clean cities. She noted that most people living in cities were breathing polluted air. Clark said cities' designs needed to be clean and sustainable. This was important for people and the planet: "We have only the . . . resources of one planet to live on, yet we live in a way which assumes that we have the resources of three, four, or more planets."

In 2014 a streetcar project in Tucson, Arizona, cost an estimated $196 million. The project laid 3.9 miles (6.3 km) of track. Systems like New York City's take years and millions of dollars to build. Despite the cost, public transportation is important to lowering emissions in cities. Public transit produces less carbon dioxide per traveler. Air quality gets better.

Public transportation takes cars off of roads. It makes city traffic lighter. Pedestrians and bikers are safer when more people use public transit.

FURTHER EVIDENCE

Chapter Three explains how the way cities are designed affects carbon emissions. What was one of the main points of this chapter? What evidence is included to support this point? Read the article at the website below. Does the information on the website support the main point of the chapter? Does it present new evidence?

NATIONAL GEOGRAPHIC: SMOG

abdocorelibrary.com/stopping-climate-change

RENEWABLE ENERGY

One of the key steps to stopping climate change is using renewable energy. Renewable energy comes from natural sources such as water, wind, or sunlight. Renewable energy does not get used up. Nonrenewable sources such as fossil fuels take millions of years to form. Humans use these fuels faster than the fuels are created.

HYDROPOWER

Hydropower comes from the movement of water. Water turns a turbine. This action

The Hoover Dam is one of the United States' largest sources of hydropower.

produces energy in a motor. The motor stores the energy as electricity in a battery. Hydropower can make energy from river currents or ocean waves. These sources of energy are consistent. A river's current will continue to flow every day.

People have used hydropower for hundreds of years. Before electricity, people used waterpower to turn mills. The mills would grind grain to make flour. Today, hydroelectric plants can be huge. Some companies build dams. Dams control how much water flows through the system. However, dams change the way rivers flow. A dam blocks the flow of water. It creates a lake. Water from the lake flows through the system slowly. Large hydroelectric systems can have a negative effect on the environment. For example, fish cannot travel upstream to lay their eggs.

Companies are starting to design new hydroelectric systems. The new systems are smaller and more efficient. They do not always need a dam in order to produce enough energy. Scientists are also exploring using ocean currents and ocean waves. Turbines would be placed underwater. However, scientists do not yet fully understand how marine life would be affected.

WIND ENERGY

Like hydropower, wind energy creates electricity through spinning turbines. Wind makes the rotors of a turbine spin. Wind turbines can be large or small. Large turbines are often part of wind farms. These farms produce electricity.

TRAVEL-SIZE POWER

Renewable energy devices are getting smaller every year. In 2015 a German company created a portable hydroelectric generator. Blue Freedom created a turbine and battery that nest together in a single disk. The turbine is only 5 inches (12 cm) wide. It easily fits in a backpack. Campers or hikers can put the turbine in a stream. Its battery can charge phones and cameras. Hydropower and other renewables do not have to be large to make a big impact.

PERSPECTIVES

OCEAN BREEZES

Wind is strongest offshore. There are no trees or hills to block it. Europe and China are leaders in offshore wind energy. The International Energy Agency (IEA) says wind energy could create enough electricity to power the world. In a 2019 statement, the IEA wrote, "Offshore wind has the potential to generate more . . . than 18 times global electricity demand today." In 2019 offshore wind energy made just 0.3 percent of the world's electricity.

Wind farms help farmers. Farmers in rural areas can sell part of their land to wind energy companies. Land that cannot be used for growing crops gets an important role. Money goes back into rural communities.

But wind turbines are loud. Additionally, turbines are not attractive to people who want to enjoy nature. There have been few studies about how wind farms disrupt wildlife. Many people do not want to live near wind farms. It is important for turbines to become quieter. One way that companies are making quieter turbines is by changing

Solar panels are expensive to install, but they can lower a home's carbon footprint.

the rotors. The faster the blades spin, the more noise is produced. One company in Europe attaches pieces of plastic to each blade's trailing edge. The pieces have a jagged edge. The jagged edge changes how the air flows off the blade. It makes the turbine quieter.

SOLAR ENERGY

Solar energy uses light to create electricity. Light from the sun strikes solar panels. These panels turn the light into electricity. That electricity can be used to power things. It also can be stored in a battery so it can be used later. The energy can be used the same as electricity from fossil fuels.

Solar panels can be fitted onto most buildings. They can be on rooftops or on parking garages. Some organizations install many solar panels on the roofs of their buildings. The organizations can sell the energy to electric companies. The electric companies can use that solar power for homes in the area. People can also install solar panels on their own homes.

Energy from solar panels is not the most efficient form of renewable energy. However, it can be used in more places than hydropower or wind energy. People can install one or many solar panels.

There are a few downsides to solar energy. Installing solar panels is expensive. Governments give rebates to people who install solar panels. However, the cost of solar panels for a home is still several thousand dollars. The panels do not work as well on cloudy days or when covered in snow. The batteries available to most people do not store a large amount of energy. Companies are working to improve battery storage and

GLOBAL SOLAR POWER PRODUCTION

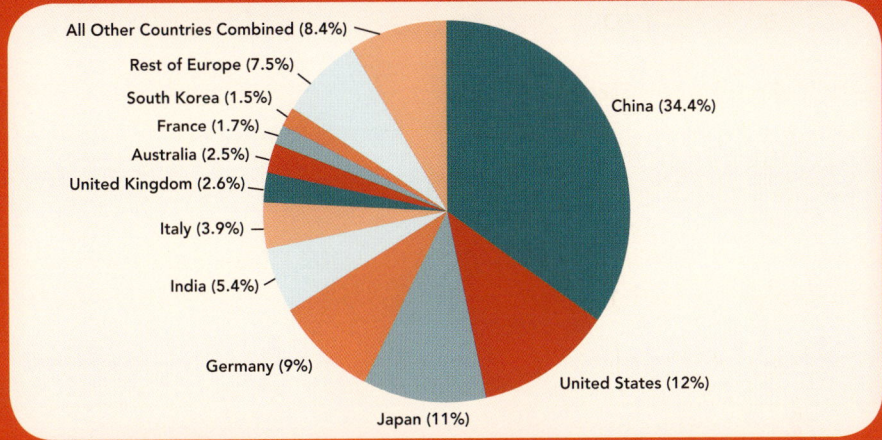

Global Solar Power Production pie chart:

- China (34.4%)
- United States (12%)
- Japan (11%)
- Germany (9%)
- India (5.4%)
- Italy (3.9%)
- United Kingdom (2.6%)
- Australia (2.5%)
- France (1.7%)
- South Korea (1.5%)
- Rest of Europe (7.5%)
- All Other Countries Combined (8.4%)

Solar panels each have a certain amount of power they can produce. This chart shows the percentage of total solar panel power each country had installed through 2018. What do you notice about the chart? How does it help you better understand the text?

moving gathered energy to consumers. This includes building new electricity grids between storage facilities and towns.

NUCLEAR ENERGY

Nuclear energy is another way to generate electricity. It is not renewable. But it also does not produce carbon emissions. In nuclear energy, atoms of an element called uranium are split apart. This splitting process is called nuclear fission. It releases energy in the form of heat. The heat boils water and creates steam. Steam turns turbines to create electricity.

Nuclear energy does not produce carbon dioxide. However, uranium is radioactive. That means it gives off a dangerous form of energy called radiation. Radiation can harm living things. The unused uranium is called nuclear waste. It must be stored safely so it doesn't hurt people. Additionally, if an accident happens at the power plant, radioactive material can explode. People can become ill. It can harm the environment.

For these reasons, some argue it is better to use renewable energy. As more people choose renewable energy, these energy sources will become more efficient and less expensive. Renewable energy, combined with other actions, can help reduce greenhouse gas emissions. Slowing and perhaps even stopping climate change will take time, money, and persistence. But it will save the environment and life as we know it.

STRAIGHT TO THE
SOURCE

Paula Garcia is an energy expert. In a 2018 interview for the Union of Concerned Scientists, she explained how renewable energy is becoming more efficient:

In terms of integrating large amounts of renewable energy like wind and solar because we always hear, "Oh, but wind is not always blowing, and sun is not always shining." Well, guess what. We have energy storage, and we can store that energy that is being produced with solar energy or wind energy. And one of the largest, if not the largest project, is located in Australia. It's [a] 100-megawatt facility, and in California there is a proposal to develop a project that is 3 times as big as that project in Australia.

So, it means that we will be able to integrate much more solar to the system through this kind of [approach].

Source: "Getting Excited about Energy: Expanding Renewables in the US." *Union of Concerned Scientists*, 21 Aug. 2018, ucsusa.org. Accessed 16 Dec. 2019.

WHAT'S THE BIG IDEA?

Take a close look at this passage. What is the main point the speaker is making about renewable energy? Name two or three supporting details.

FAST FACTS

- Climate change is causing Earth to warm. Because of this warming, animals' habitats are changing, and storms are becoming more severe.

- The burning of fossil fuels adds greenhouse gases to the air, causing climate change.

- Many people are working to slow the effects of climate change.

- Carbon taxes are taxes on carbon emissions. They can work because people change their energy use and sources to avoid paying the tax.

- Adding green spaces to cities means making areas with lots of plants. The plants take in carbon, reducing the amount of carbon in the air.

- If many people in an area use public transit, they can have a smaller carbon footprint than if they each used their own personal cars. Also, electric public transit vehicles can reduce greenhouse gas emissions if the energy comes from renewable energy sources.

- Renewable energy includes hydropower, wind energy, and solar energy. Water flow, wind, and sunlight are energy sources that can be used again and again without getting used up.

- Nuclear energy does not produce greenhouse gas emissions. But it is radioactive, making it very dangerous to living things if there is an accident.

STOP AND
THINK

Tell the Tale

Chapter One of this book discusses one person's decision to use public transportation instead of driving to work in her car. Imagine you are considering making a similar decision. Write 200 words about how your life might change if you used only public transportation. What other choices could you make to use less fossil fuels?

Surprise Me

Chapter Three discusses changing how people design cities to be more environmentally friendly. After reading this book, what two or three facts about climate change did you find most surprising? Write a few sentences about each fact. Why did you find each fact surprising?

Dig Deeper

After reading this book, what questions do you still have about attempting to stop climate change? With an adult's help, find a few reliable sources that can help you answer your questions. Write a paragraph about what you learned.

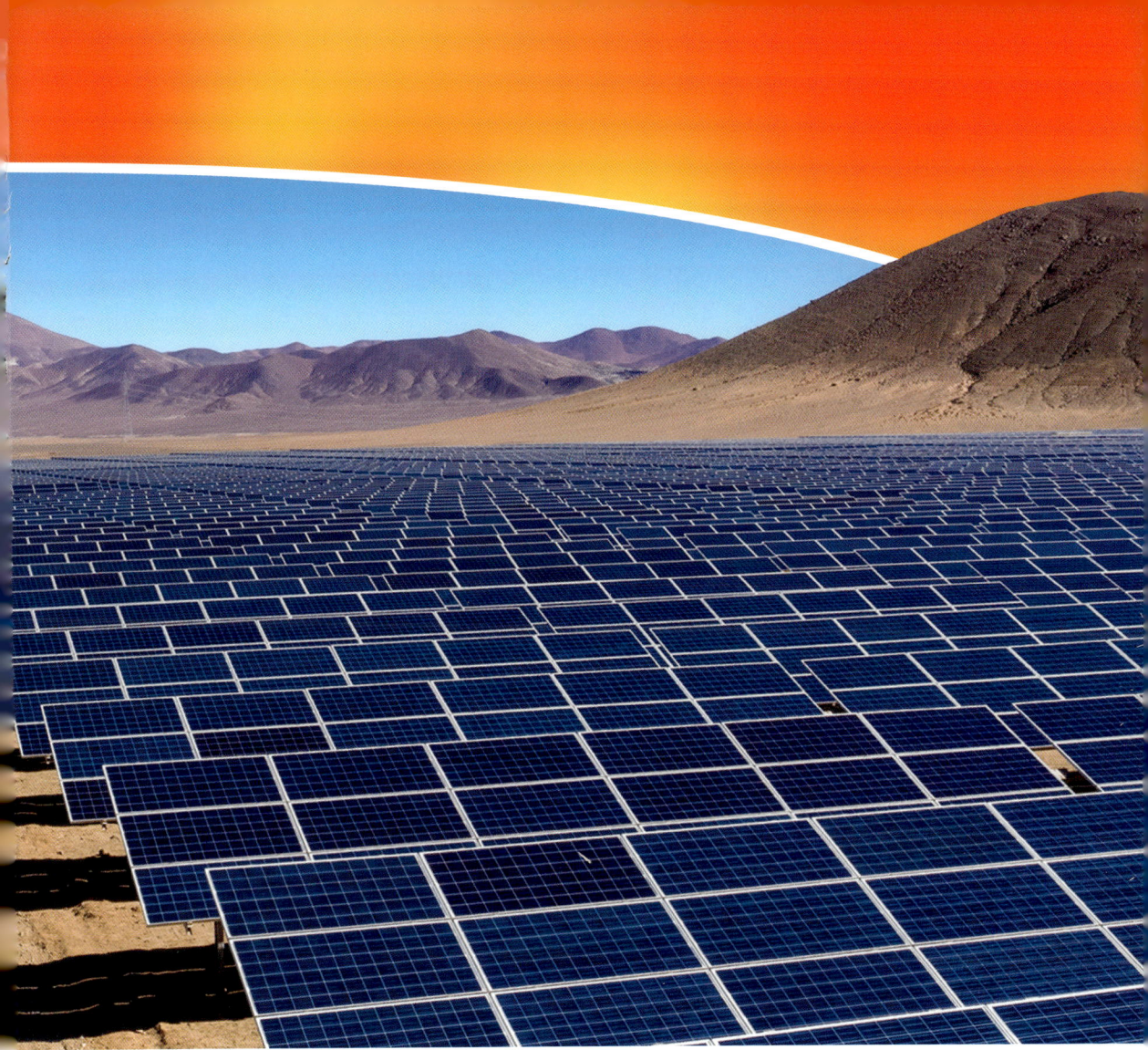

You Are There

This book discusses how vertical forests can help slow climate change. Imagine you are in Milan, Italy. Write a letter to your friends about your experience in the vertical forests. What do you notice about the buildings? Be sure to add plenty of detail to your notes.

GLOSSARY

compost
food scraps from a kitchen that break down easily into dirt and fertilizer

emission
gas that is released by an object or action

greenhouse gas
a gas such as methane or carbon dioxide that traps heat in the atmosphere

hybrid
using both renewable and nonrenewable energy

idle
to remain on or running while not moving

incentive
a reward for changing a behavior

pedestrian
a person traveling on foot

rebate
money given back to consumers

rotor
the spinning blade on a turbine

turbine
a machine that produces energy from spinning blades

ONLINE RESOURCES

To learn more about stopping climate change, visit our free resource websites below.

Core Library
CONNECTION
FREE! COMMON CORE MULTIMEDIA RESOURCES

Visit **abdocorelibrary.com** or scan this QR code for free Common Core resources for teachers and students, including vetted activities, multimedia, and booklinks, for deeper subject comprehension.

Booklinks
NONFICTION NETWORK
FREE! ONLINE NONFICTION RESOURCES

Visit **abdobooklinks.com** or scan this QR code for free additional online weblinks for further learning. These links are routinely monitored and updated to provide the most current information available.

LEARN MORE

Huddleston, Emma. *Adapting to Climate Change*. Abdo Publishing, 2021.

Murray, Laura K. *Ocean Energy*. Abdo Publishing, 2017.

INDEX

About the Author

Martha London writes books for young readers. When she isn't writing, you can find her hiking in the woods.